BUILD 10 BOLD BRIDGES WITH STEM

by Chelsey Luciow

CAPSTONE PRESS
a capstone imprint

Dabble Lab is published by Capstone Press, an imprint of Capstone.
1710 Roe Crest Drive, North Mankato, Minnesota 56003
capstonepub.com

Copyright © 2025 by Capstone. All rights reserved. No part of this publication may be reproduced in whole or in part, or stored in a retrieval system, or transmitted in any form or by any means, electronic, mechanical, photocopying, recording, or otherwise, without written permission of the publisher.

Library of Congress Cataloging-in-Publication Data is available on the Library of Congress website.
ISBN: 9781669086437 (hardcover)
ISBN: 9781669086482 (ebook PDF)

Summary: Engineers know there's always more than one way to solve a problem. Use your STEM smarts to brainstorm, design, and build a bridge in 10 different ways. Each project will introduce you to the STEM concepts at the heart of bridge building.

Image Credits: Adobe Stock: dimagroshev, 5 (research), drawlab19, 5 (create), katarinalas, 5 (test, improve), notkoo2008, 5 (plan), sdecoret, 5 (ask, brainstorm); Mighty Media, Inc.: project photos; Shutterstock: Tatiana Popova, 14 (cars)
Design Elements: Adobe Stock: drawlab19, sidmay; Shutterstock: N.Savranska

Editorial Credits
Editor: Jessica Rusick
Designer: Denise Hamernik

Any additional websites and resources referenced in this book are not maintained, authorized, or sponsored by Capstone. All product and company names are trademarks™ or registered® trademarks of their respective holders.

The publisher and the author shall not be liable for any damages allegedly arising from the information in this book, and they specifically disclaim any liability from the use or application of any of the contents of this book.

Printed and bound in China. 6096

TABLE OF CONTENTS

Build Bridges with STEM 4
Decked-Out Bridges. 6
Trusty Truss Bridge 10
Spaghetti Structure. 12
Crafty Clothespin Bridge 14
Rubber Band Roadway 16
Dynamic Drawbridge 18
Straw Suspension Bridge 20
Amazing Arch Bridge 22
Legendary Lift Bridge 24
Fairy Bridge 28
    Read More 32
    Internet Sites 32
    About the Author 32

BUILD BRIDGES WITH STEM

Think about the last time you crossed a bridge. Was it large or small? Did it carry people or cars? Engineers build strong and stable bridges using science, technology, engineering, and math (STEM). You can too!

BASIC SUPPLIES

- ✓ cardboard
- ✓ duct tape
- ✓ hot glue gun and glue sticks
- ✓ markers
- ✓ rubber bands
- ✓ ruler
- ✓ scissors
- ✓ string
- ✓ toothpicks

BUILDING TIPS

Set yourself up for success! Read through the materials and instructions before starting a project.

Let your creativity shine! Put your own stamp on these projects. Make changes or try something new! Is there a material you can't find? Think of ways to adapt the project using the items you have.

Ask first! Get permission to use any materials you need.

Safety first! Ask an adult for help with projects that require sharp or hot tools.

Clean up! When you're finished building, make sure to put away any supplies you took out and clean up your space.

Keep trying! Don't worry if your bridge breaks or falls over. Think about how to improve your bridge. Then try again!

ENGINEERING DESIGN PROCESS

The engineering design process helps engineering teams think through problems and design devices to solve them. Use this process as you build the bridges in this book!

ASK
What problem needs to be solved?

RESEARCH
How have people tried to solve the problem in the past? What materials or technologies did they use?

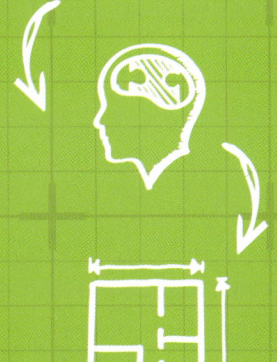

BRAINSTORM
What are potential new ways to solve the problem? Don't be afraid to get creative and imagine wild, wacky ideas!

PLAN
Which solution seems the most promising? How will you turn this idea into a real design?

CREATE
Build a prototype, or early version, of your design.

TEST
Does the prototype solve the problem? How could the prototype be improved?

IMPROVE
Redesign and improve your prototype. Repeat the process to solve new problems!

DECKED-OUT BRIDGES

Try different techniques to build bridges out of playing cards.

MATERIALS

- ✓ marker
- ✓ ruler
- ✓ deck of old playing cards
- ✓ scissors

Making the Bridge's Deck

STEP 1

Draw two lines on one long side of a card. Make them about 0.5 inches (1.3 centimeters) long and a little more than 1 inch (2.5 cm) apart. Cut each line to make a tab on the card.

STEP 2

Slide another card above the tab of the first card. Trace the two tab lines onto the new card.

STEP 3

Slide the cards apart and cut the new lines you drew. Then fit the cards together so they interlock. Draw and cut lines on the other long side of the newer card.

STEP 4

Repeat steps 2 and 3 with at least 10 more cards to make an interlocking card chain. This card chain is the bridge's deck. The deck of a bridge is the part that people and cars move across.

Project continues on the next page.

7

Adding Supports: Technique 1

STEP 1

Supports help a bridge stand upright. They bear the weight of the deck so a bridge doesn't collapse. Repeat steps 1 through 4 of Making the Bridge's Deck to make another card chain. But this time, make the cuts on the short sides of the cards and use fewer cards.

STEP 2

Lay the short card chain flat. Arch the deck over it. Make horizontal cuts in each card where the two chains meet. The cuts should be in opposite directions so the cards can interlock. Slide the two ends of the chains together at the cuts.

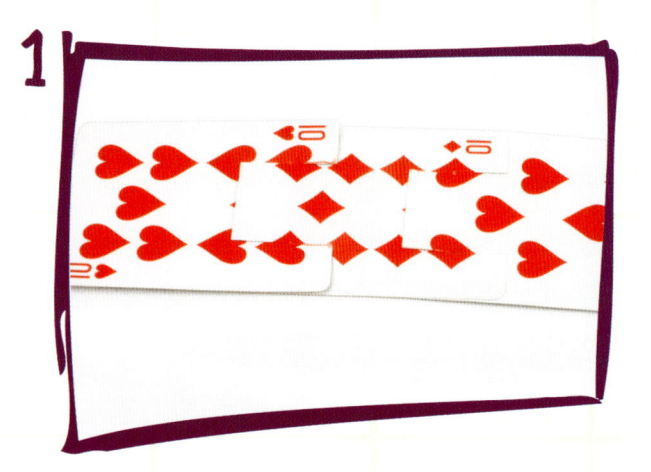

STEM Break

When a bridge's deck bears weight, the weight both pulls the bridge in opposite directions and compresses it down. Strong bridges are designed to spread these forces equally through the supports so no one part of the bridge bears too much. Which type of support worked best in your card bridges? How could you make the bridges stronger?

Adding Supports: Technique 2

STEP 1
Draw a line from the center of a card's long side to the middle of the card. Cut on the line. Repeat with another card.

STEP 2
Interlock the two cards along the cuts to make an X shape.

STEP 3
Repeat steps 1 and 2 to make another X shape. Stand the shapes upright and close together.

STEP 4
Lay the bridge's deck on top of the supports.

STEP 5
Experiment by moving the supports farther apart or turning them different ways. You can also make the deck longer or shorter.

TRUSTY TRUSS BRIDGE

Harness the power of triangles to build a surprisingly strong structure.

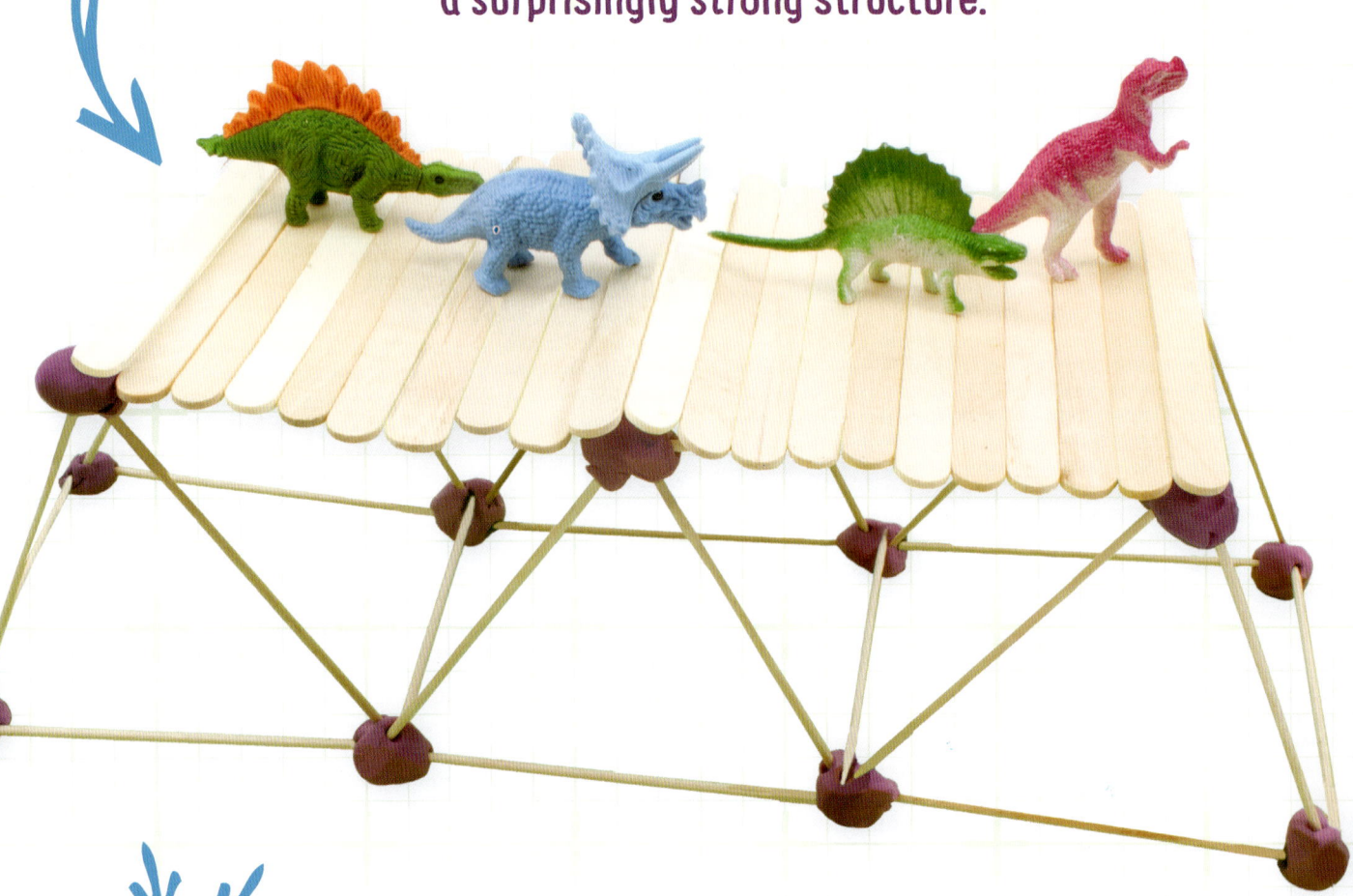

MATERIALS
- ✓ air-dry clay
- ✓ toothpicks
- ✓ craft sticks
- ✓ duct tape

STEP 1
Roll three small clay balls. Use the balls to attach three toothpicks together in a triangle. Position the triangle so one point faces up.

STEP 2

Form another triangle with three toothpicks and two balls. Attach the corner with no ball to the bottom right corner of the first triangle. You should now have two triangles connected by one corner.

STEP 3

Repeat step 2 to connect a third triangle to the second triangle. Make the triangles one closed trapezoid shape by connecting the top points with toothpicks.

STEP 4

Repeat steps 1 through 3 to make another trapezoid.

STEP 5

Stand the trapezoids upright so they are parallel to each other. Connect the trapezoids together at the top and bottom with toothpicks.

STEP 6

Lay craft sticks across the top of the structure to see how many it takes to fill the top of the bridge. Remove the craft sticks and duct tape them together to make one long piece for the bridge's deck. Lay the deck on top of the bridge, duct tape side down.

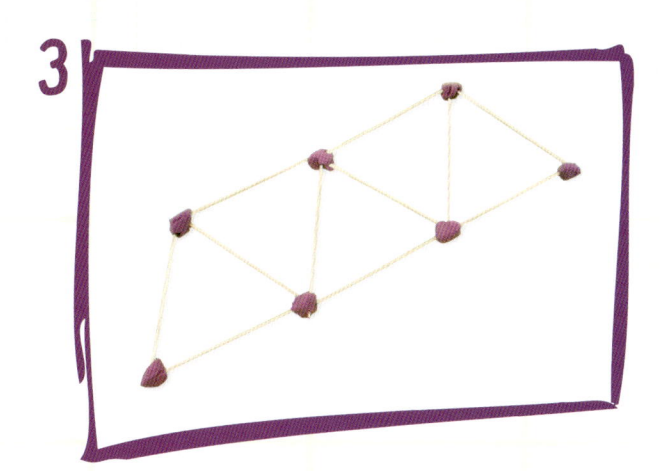

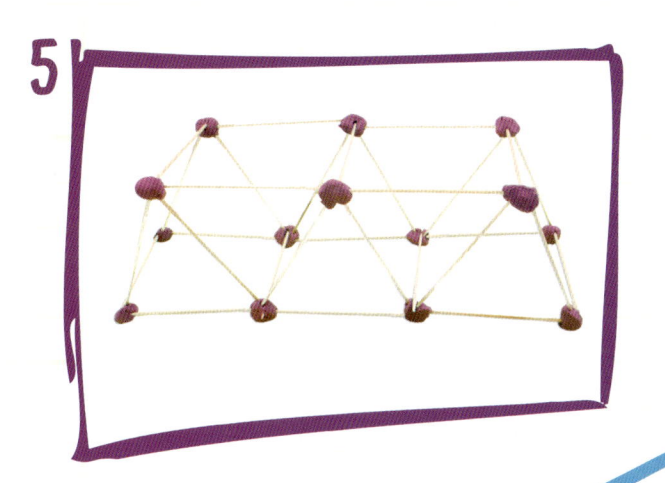

STEM Break

A triangle is a strong shape for building. Forces are distributed in a balanced way through its sides and angles. This means triangles don't collapse easily, even under lots of weight. Many bridges are made with strings of triangles called trusses for this reason.

SPAGHETTI STRUCTURE

Connect bundles of noodles together to explore this project's bridge-building pasta-bilities.

MATERIALS
- ✓ dry spaghetti noodles
- ✓ mini rubber bands

STEP 1
Create bundles by banding eight to ten noodles together. Make at least 18 long bundles with whole noodles and at least 18 short bundles with noodles broken in half.

12

STEP 2

Lay three short bundles on top of each other in a triangle. The corners should overlap. Connect the triangle at the corners with rubber bands. Repeat to make a second identical triangle.

STEP 3

Stand the triangles from step 2 upright. Lay three long spaghetti bundles between the corners of the triangles to create the bridge's 3D structure.

STEP 4

Repeat steps 2 and 3 to make two more structures. Lay long noodle bundles across the structures to connect them together. Rubber band the bundles in place. Then use more noodle bundles or individual noodles to make the bridge's deck.

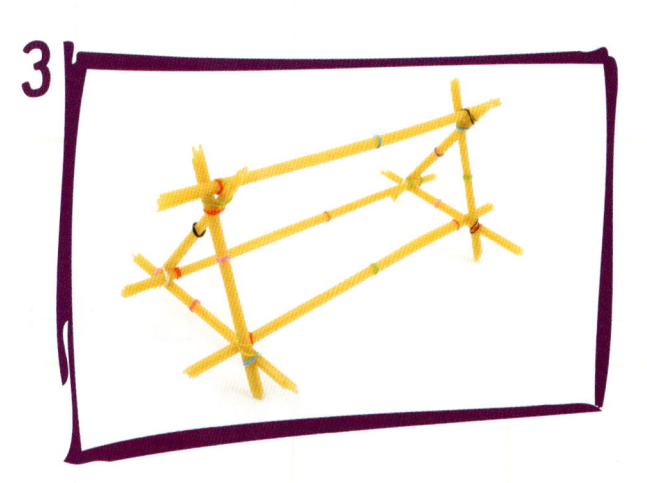

STEM Break

One noodle can't withstand much downward force. Noodle bundles are strong building materials because forces are spread between many noodles. There is not as much force on any one noodle.

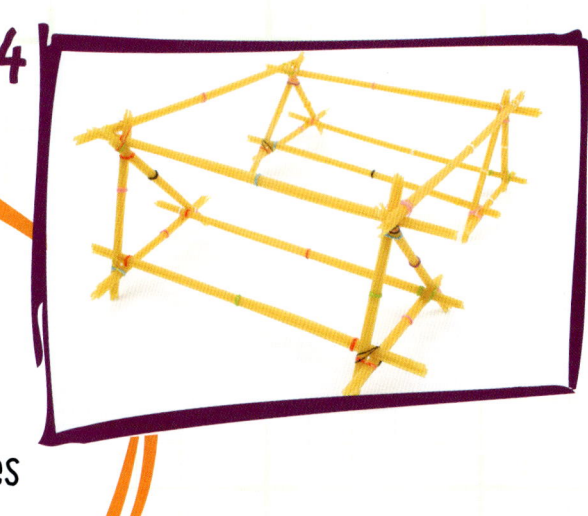

13

CRAFTY CLOTHESPIN BRIDGE

Explore the properties of a beam bridge with this simple clothespin creation.

MATERIALS

- ✓ 4 spring clothespins
- ✓ 2 straws
- ✓ craft sticks
- ✓ masking tape
- ✓ small toy cars (optional)

14

STEP 1
Stand two clothespins upright. Clip a straw between them. Repeat with the other clothespins and straw. Stand the two clothespin structures next to and parallel to each other.

STEP 2
Lay the craft sticks between the two straws. Experiment with the distance between the straws.

STEP 3
Measure how many craft stick ends fit between the clothespins at one end of the bridge. Tape that many craft sticks together side by side to make a ramp. Repeat to make a second ramp.

STEP 4
Lean the ramps on either side of the bridge. Test the bridge with the weight of small toy cars.

STEM Break

At its most basic, a beam bridge is a horizontal beam laid between two vertical towers. The towers support the bridge's weight. In this project, the straws transfer the weight of the craft sticks and cars to the clothespins.

15

RUBBER BAND ROADWAY

Use rubber bands to create a sturdy beam bridge.

MATERIALS

- ✓ 12-egg egg carton
- ✓ 12 colored pencils
- ✓ 12 rubber bands
- ✓ ruler
- ✓ toy car

16

STEP 1
Open the egg carton. Poke the pointed ends of two colored pencils up through the underside of the first two egg slots. Wrap a rubber band twice around the pencils to hold them together.

STEP 2
Repeat step 1 with the remaining pencils and slots. Then close the egg carton so it is upside down and the bottoms of the pencils are sticking up. Straighten the pencils as needed.

STEP 3
Connect each pair of pencils with a rubber band. Wrap the rubber band twice around one pencil. Then stretch the rubber band and wrap it twice around the other pencil.

STEP 4
Lay a ruler on the rubber bands. Adjust the heights of the rubber bands if needed to make the bridge level. Test the bridge with a toy car.

STEM Break

Weight from the ruler and the car causes the rubber bands to stretch, creating tension. This tension creates a force that pushes back up on the bridge to support it.

DYNAMIC DRAWBRIDGE

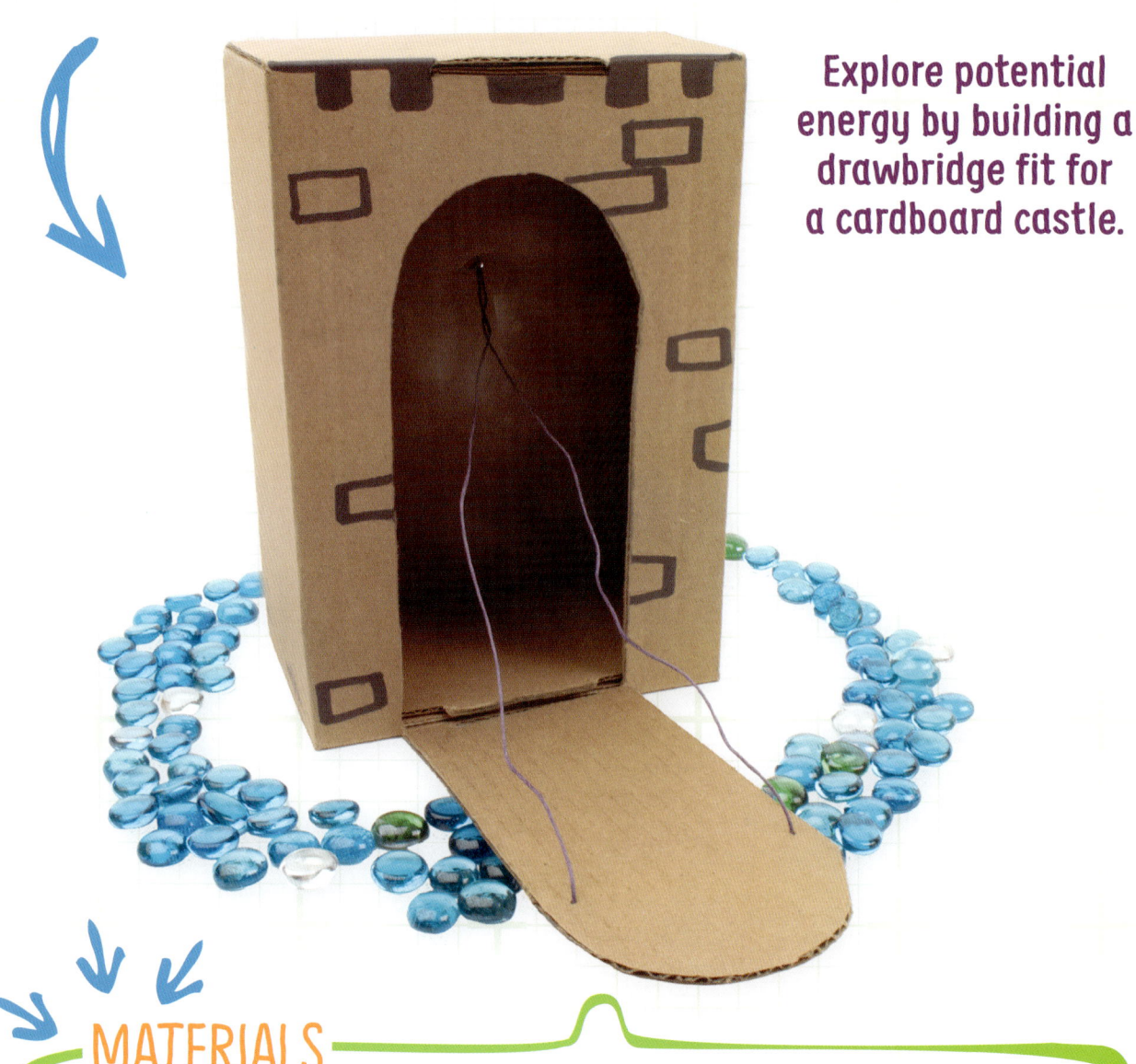

Explore potential energy by building a drawbridge fit for a cardboard castle.

MATERIALS

- ✓ large roll of tape
- ✓ small cardboard box
- ✓ marker
- ✓ ruler
- ✓ craft knife
- ✓ toothpick
- ✓ string
- ✓ scissors
- ✓ spring clothespin
- ✓ weights such as rocks or coins
- ✓ hot glue gun and glue sticks

STEP 1
Make the bridge's door. Place the roll of tape on the front of the box. Trace the top curve and use the ruler to draw straight lines down from the curve. Cut out the door on all sides except the bottom.

STEP 2
Open the door. Poke a small hole on each side of the door near the edge.

STEP 3
Cut two strings that can reach from the door holes to the back of the box when the door is open. Feed one string through one door hole from the outside in and tie a knot at the front of the door. Repeat with the other string and hole.

STEP 4
Poke a hole in the center of the back of the box and feed both strings through it. Knot the strings together.

STEP 5
Pull the knotted string to close the door. Use a clothespin to hold the string in place.

STEP 6
Hot glue the weights to the door's front to help the door open. Remove the clothespin to open the drawbridge. Pull the string and clip it in place to close the drawbridge.

STEM Break

The closed drawbridge has potential energy. This is energy stored in objects because of their positions. When you unclip the clothespin, the weights pull on the door, turning the potential energy into kinetic energy, or the energy of motion.

STRAW SUSPENSION BRIDGE

Learn about the properties of suspension bridges with this string-and-straw creation.

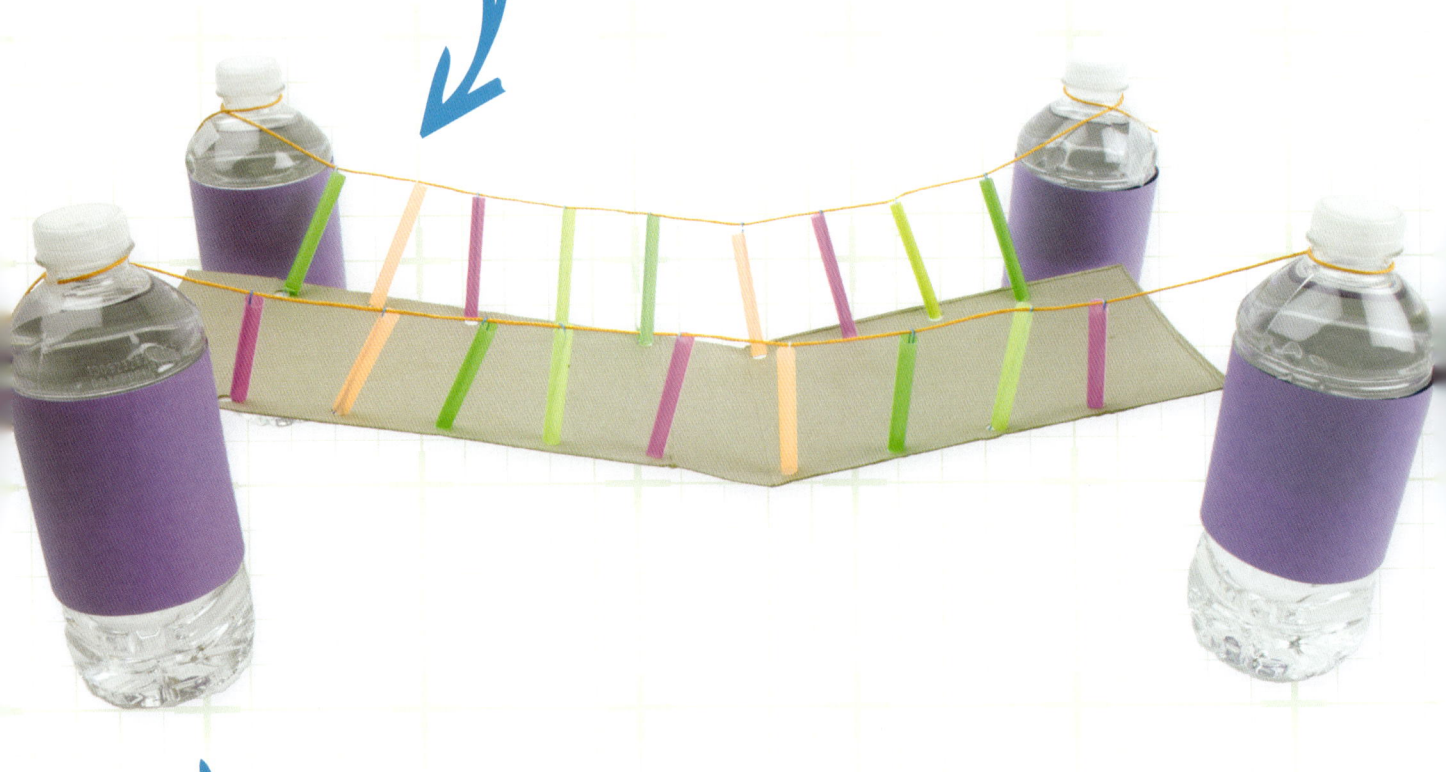

MATERIALS

- thin cardboard (cereal box or similar)
- hole punch
- scissors
- straws
- thick and thin string
- tape
- 4 full, identical water bottles

STEP 1
Cut a rectangle out of the cardboard. Punch evenly spaced holes on each long side.

STEP 2
Cut straws into 1-inch (2.5-cm) pieces. There should be one piece for each hole in the cardboard.

STEP 3
Cut one thin string and one thick string twice as long as the cardboard. Lay the thick string horizontally on the cardboard.

STEP 4
Thread the thin string up through the first hole and through a straw piece. Loop the thin string over the thick string and back down through the same straw and hole. Repeat this process on one side with the other holes. Tape the leftover thin string to the bottom of the cardboard.

STEP 5
Repeat steps 3 and 4 on the other side.

STEP 6
Suspend the bridge by tying each end of thick string to the neck of a water bottle.

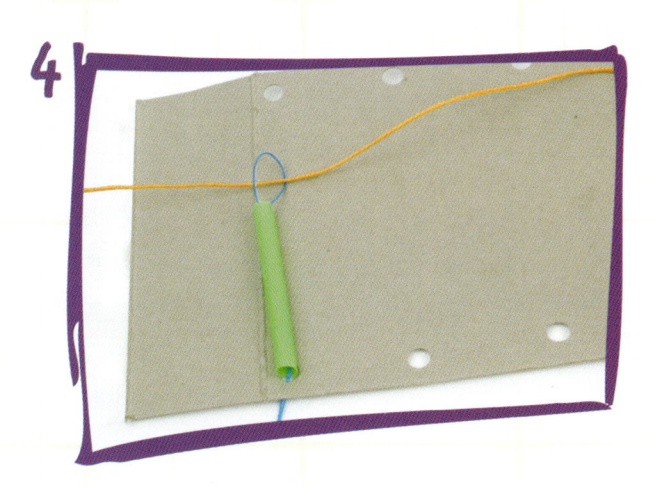

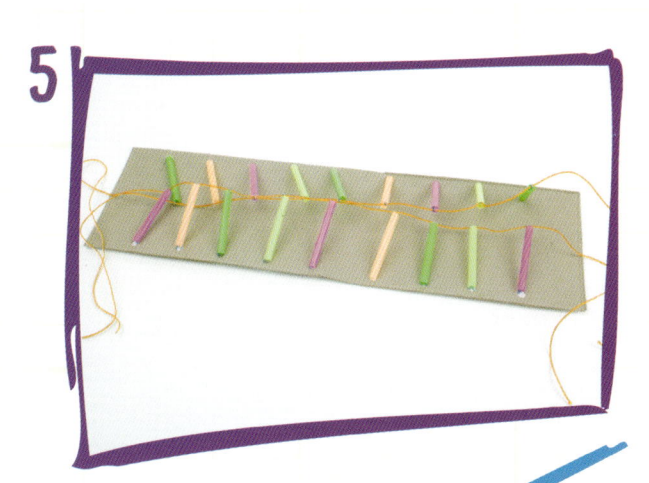

STEM Break

A suspension bridge's deck hangs from cables connected to large towers. The weight of the deck and anything on it pulls the cables. The tension in the cables compresses the towers. The towers transfer this compression into the ground.

AMAZING ARCH BRIDGE

Cut out foam blocks to build a self-supporting arch bridge.

MATERIALS

- large, thick block of cushion foam
- marker
- scissors
- paint and paintbrush (optional)
- small toys (optional)

STEP 1
Draw an arch shape on the foam. Cut it out.

STEP 2
Use the marker to divide the arch into nine roughly even-sized pieces. The pieces should become more wedge-shaped as they get closer to the arch's center.

STEP 3
Cut the pieces apart. Paint them if you'd like. These are the bridge's building blocks.

STEP 4
Assemble the arch upright. Set down the base block on one side. Then set down the other side's base block. Keep stacking the blocks like this, alternating between sides, until only the center block remains. You or a partner may have to hold the sides to keep them from falling.

STEP 5
Place the center block at the top to connect the two sides of the arch. The arch bridge should now stay standing on its own. If you'd like, try putting small toys on top to see how much weight it can hold.

STEM Break

Some bridges are held together with nothing but their own weight! In this bridge, the center block, or keystone, helps transfer weight through the other blocks and then into the ground.

23

LEGENDARY LIFT BRIDGE

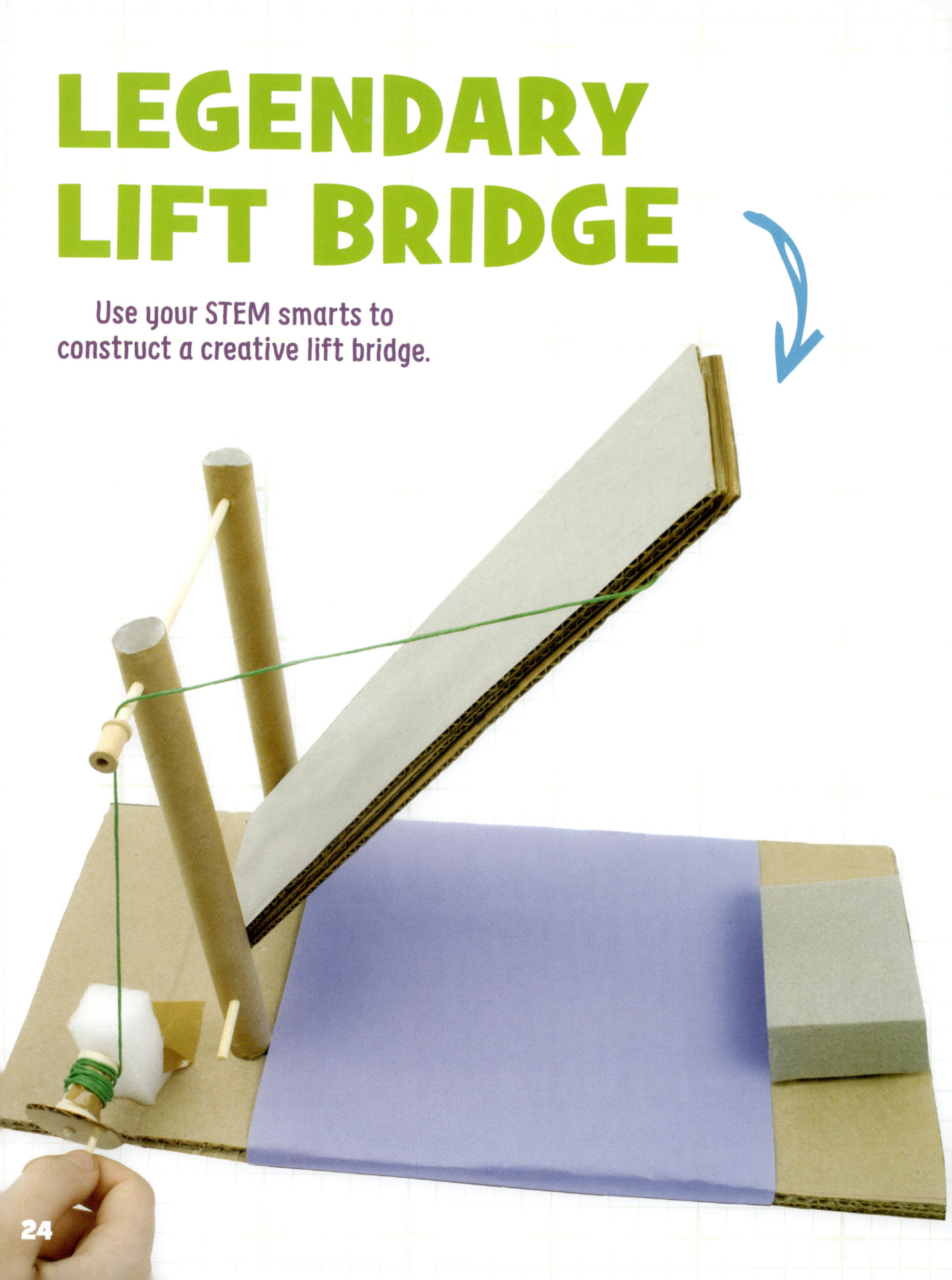

Use your STEM smarts to construct a creative lift bridge.

MATERIALS

- cardboard
- scissors or craft knife
- ruler
- hot glue gun and glue sticks
- marker
- 2 cardboard tubes
- three 12-inch (30.5-cm) wooden skewers
- straw
- small packing foam brick
- 2 spools (1 large, 1 small)
- string
- duct tape
- decorating supplies (optional)

STEP 1

Cut two rectangles out of cardboard. They should be about 11 by 6 inches (27.9 by 15.2 cm). Hot glue them together. This is the bridge's base.

STEP 2

Cut a cardboard rectangle about 8.25 inches (21 cm) by 3 inches (7.6 cm). This is the bridge's deck. Cut three cardboard rectangles that are 3 inches (7.6 cm) by 1 inch (2.5 cm). Hot glue the three small rectangles together to make the bridge's landing.

STEP 3

Hot glue the landing along a short edge of the base. Lay the deck on top of the landing so it extends over the base. Make two marks on the base, one on each side of the deck's end opposite the landing. The marks are where the bridge's pillars will go.

Project continues on the next page.

STEP 4
Hot glue a cardboard tube upright on each mark. These are the bridge's pillars.

STEP 5
Push a wooden skewer horizontally through the pillars. The skewer should be about 1 inch (2.5 cm) above the base.

STEP 6
Cut a straw the width of the deck. Hot glue the straw onto the bottom of the deck at one end.

STEP 7
Pull the skewer out. Push the skewer into the first pillar, through the straw from step 6, and through the second pillar. The deck should now extend from the pillars and rest on top of the landing.

STEP 8
Push a second skewer through the pillars 1 inch (2.5 cm) from the top.

STEP 9
Make the bridge's wheel and axle system. Cut a cardboard circle about 1.5 inches (3.8 cm) in diameter for the wheel. Poke a skewer into the wheel's center for the axle and hot glue it in place. Cut 1 inch (2.5 cm) off the axle's end for the crank. Hot glue the crank to the wheel's edge so it sticks out opposite the axle.

STEP 10
Slide the large spool and foam brick onto the axle and hot glue them in place. Cut off any extra axle that sticks out from the foam block.

STEP 11
Cut a string to be several feet long. Duct tape one end to the bottom of the deck near the landing. Tape the other end to the large spool. Then tape the foam block to the base behind a pillar. The crank should face out.

STEP 12
Hot glue the small spool to the skewer sticking out from the top of a pillar. Stretch the loose string from the deck up and over the skewer and down to the large spool. (The small spool will keep the string from slipping off the skewer.) Wrap the extra string around the large spool. Add decorations if you'd like. Then turn the crank to raise the lift bridge.

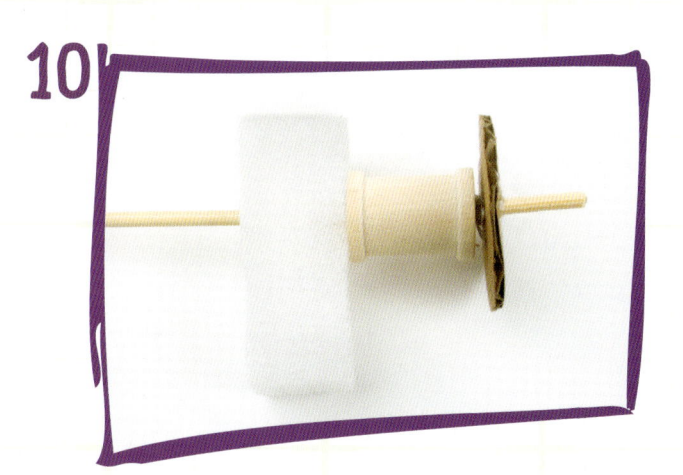

STEM Break

This lift bridge uses a crank-operated wheel and axle to raise and lower. You apply force to the wheel by turning the crank. The turning wheel applies an even greater force to the string that wraps around the axle. This force is what makes the lift bridge lift!

FAIRY BRIDGE

Use arches and triangles to build a sturdy bridge fit for a fairy wonderland.

MATERIALS

- straight and curved sticks roughly as thick as pencils
- ruler
- wire cutters
- craft wire
- hot glue gun and glue sticks
- decorating supplies (optional)

28

STEP 1

Use wire cutters to cut several sticks so they are 2.5 inches (6.4 cm) long. Line up these sticks side by side to make the deck of the bridge. The deck should be about 10 inches (25.4 cm) long.

STEP 2

Cut two pieces of wire to be about 10 inches (25.4 cm) long. Hot glue the wires next to each other across the deck of the bridge. The wires will support the deck and allow you to shape it. Let the glue dry.

STEP 3

Once the glue is dry, slowly and gently bend the bridge's deck into an arch.

Project continues on the next page.

29

STEP 4

Cut two sticks to be 5 inches (12.7 cm) long. Hot glue the sticks upright to the center of the arch on either side. The sticks should touch the ground when the arch stands upright. Let the glue dry.

STEP 5

Cut four sticks to be 2.5 inches (6.4 cm) long. Hot glue the sticks upright to each corner of the arch. These sticks should also touch the ground when the arch stands upright. Let the glue dry.

STEP 6

Hot glue four curved sticks along the arch between the upright sticks.

STEP 7

The sticks from steps 4 and 5 are posts that form the base of the bridge's railing. Hot glue four sticks in place to make the railings. Each stick should extend from the top of a long stick to the top of a short stick. Let the glue dry.

STEP 8

Cut four more sticks that are 2.5 inches (6.4 cm) long. Hot glue each one vertically between a stick from step 6 and the railing about halfway between the bridge's end and its center post. These sticks are posts that will add extra support to the railing. Let the glue dry.

STEP 9

The bridge now has five vertical posts and four gaps between those posts on each side. Measure the diagonal distance across one gap and cut two sticks that size. Hot glue the sticks in place across the gap so they form an X. Repeat to fill the other gaps with Xs. Decorate the bridge with stones, pine cones, and artificial moss if you'd like.

STEM Break

This bridge combines elements of an arch bridge and a truss bridge. Weight on top of the arch is distributed throughout its curve, giving the bridge natural strength. The crossed sticks form trusses that add extra support to the handrails.

Read More

Mattern, Joanne. *Amazing Bridges*. Mankato, MN: Black Rabbit Books, 2025.

Olson, Elsie. *Engineering Lab: Explore Structures with Art & Activities*. Minneapolis: Abdo Publishing, 2024.

Reeves, Diane Lindsey. *Do You Like Experimenting with STEM? Career Clues for Kids*. Ann Arbor, MI: Cherry Lake Publishing, 2023.

Internet Sites

The Kid Should See This: Spaghetti Bridges, a DIY Engineering Activity for Kids (and Adults)
thekidshouldseethis.com/post/spaghetti-bridges-engineering-challenge-activities-for-kids

PBS Kids for Parents: Build and Test Paper Bridges
pbs.org/parents/crafts-and-experiments/build-and-test-paper-bridges

Wonderopolis: How Long is the Longest Bridge?
wonderopolis.org/wonder/how-long-is-the-longest-bridge

About the Author

Chelsey Luciow is an artist and creator. She loves reading with kids and believes books are magical. Chelsey lives in Minneapolis with her wife, their son, and their dogs.